T5-CVH-947

SPACE BUSTERS

Space Dramas

Chris Woodford

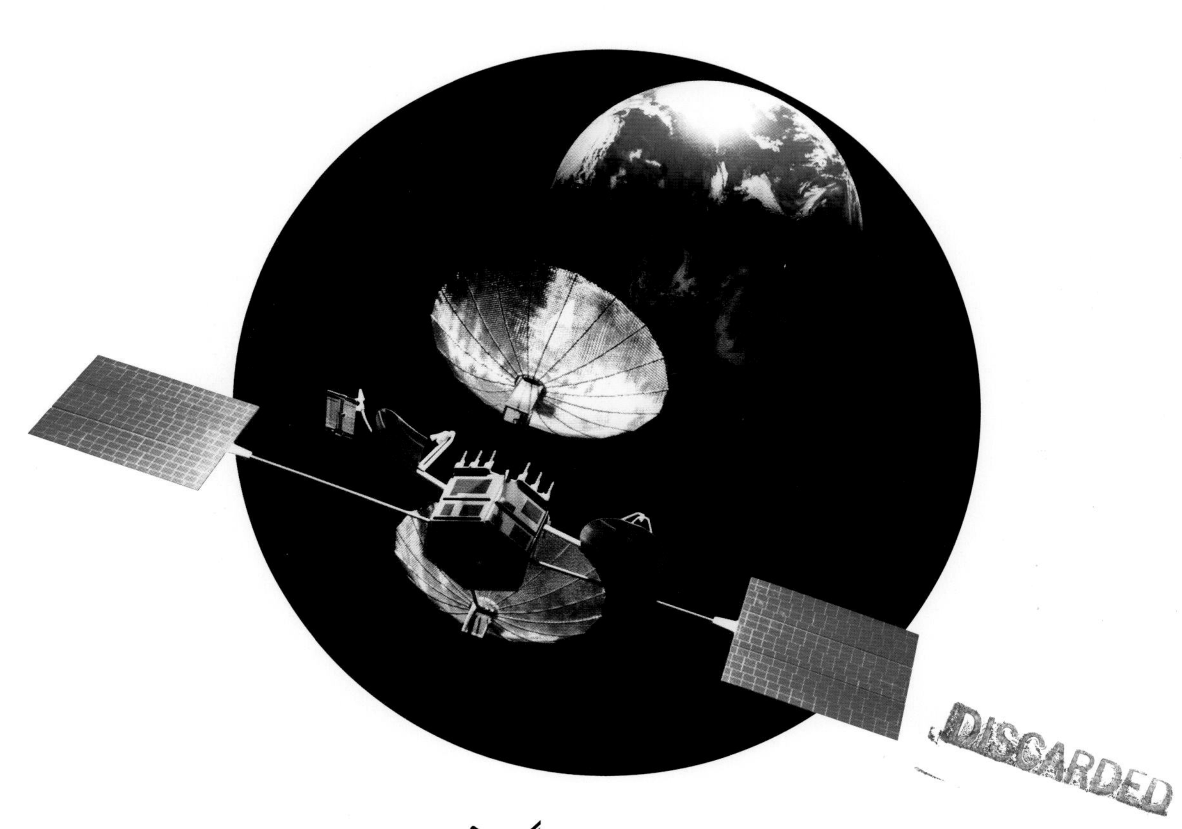

A Harcourt Company

Austin New York

www.raintreesteckvaughn.com

LOOK FOR THE RINGS OF SATURN

Look out for boxes like this with the planet Saturn in the corner. They contain extra information and amazing space-buster facts and figures.

Published by Raintree Steck-Vaughn Publishers, an imprint of Steck-Vaughn Company

Designer: Katrina ffiske
Editors: Sarah Doughty, Pam Wells
Consultant: Steve Parker

Library of Congress Cataloging-in-Publication Data
Woodford, Chris.
Space Dramas / Chris Woodford.
p. cm.—(Space busters)
Includes index.
Summary: Discusses various situations faced by astronauts in space, including dangerous dockings, daring repairs, risky rescues, and unhappy landings..

ISBN 0-7398-4850-X
1. Space vehicle accidents—Juvenile literature. [1. Space vehicle accidents. 2. Astronautics] I. Title. II. Series.

TL867.W66 2002
363.12'4—dc21 2001034943

Printed in Hong Kong.
Bound in the United States.
1 2 3 4 5 6 7 8 9 0 LB 05 04 03 02 01

Acknowledgments
We wish to thank the following individuals and organizations for their help and assistance and for supplying material in their collections: Camera Press 25 bottom (Bob Penn); Corbis 2 (Bettmann), 7 both (Roger Ressmeyer), 11 bottom (Roger Ressmeyer), 14 (NASA/Roger Ressmeyer), 20 top (Bettmann), 21 top (Bettmann), 23 top (Reuters NewMedia Inc), 26 (Reuters NewMedia Inc), 27, 31 (David Samuel Robbins); MPM Images 1, 3, 5, 6, 9, 10, 13 bottom, 17, 18, 24, 25 top, 28, 29; NASA 15; Science and Society Picture Library 19 (NASA), 22 top (NASA); Science Photo Library 11 top (European Space Agency), 12 (John Sanford), 13 top (MSSSO/ANU), 16 top (European Space Agency), 16 bottom (David Parker), 22 bottom (NASA); Topham Picturepoint 4, 8 (Adam Tanner/The Image Works), 19 top, 20 bottom, 23.

▼ **The *Ranger 7* spacecraft sent thousands of photographs back to Earth before crashing into the Moon in 1964.**

▶ **Astronauts have to make many dangerous spacewalks to build the new International Space Station.**

Contents

The Drama of Space

Have you ever felt scared going into a dark room? Imagine then how astronauts must feel when they set off in a rocket into space. The world of stars and planets is dark and distant, and it is full of danger and excitement.

A rocket launches astronauts into space with enormous power and speed. But a rocket is a complicated machine, and many different things can go wrong. Reaching space safely is only the start of the astronauts' mission, a journey into space.

◀ Many films, such as *Mission to Mars,* have been made about space and what it might be like.

Astronauts often have to do a number of dangerous jobs in space. They are a long way from Earth if anything goes wrong. Sometimes it can be difficult to get them back to Earth safely. Fortunately, most of the astronauts who travel into space live to tell their dramatic tales!

DREAMY DRAMAS

In the seventeenth century, a German astronomer named Johannes Kepler tried to imagine what traveling into space would be like. He thought people might go to the Moon on a bridge made out of a shadow and find it full of scary creatures like huge snakes with wings. Fortunately, nothing like this was found when astronauts reached the Moon in 1969!

◀ Space can seem a scary place when you are millions of miles from home.

COUNTDOWN TO DANGER

Blastoff is a very carefully controlled explosion.

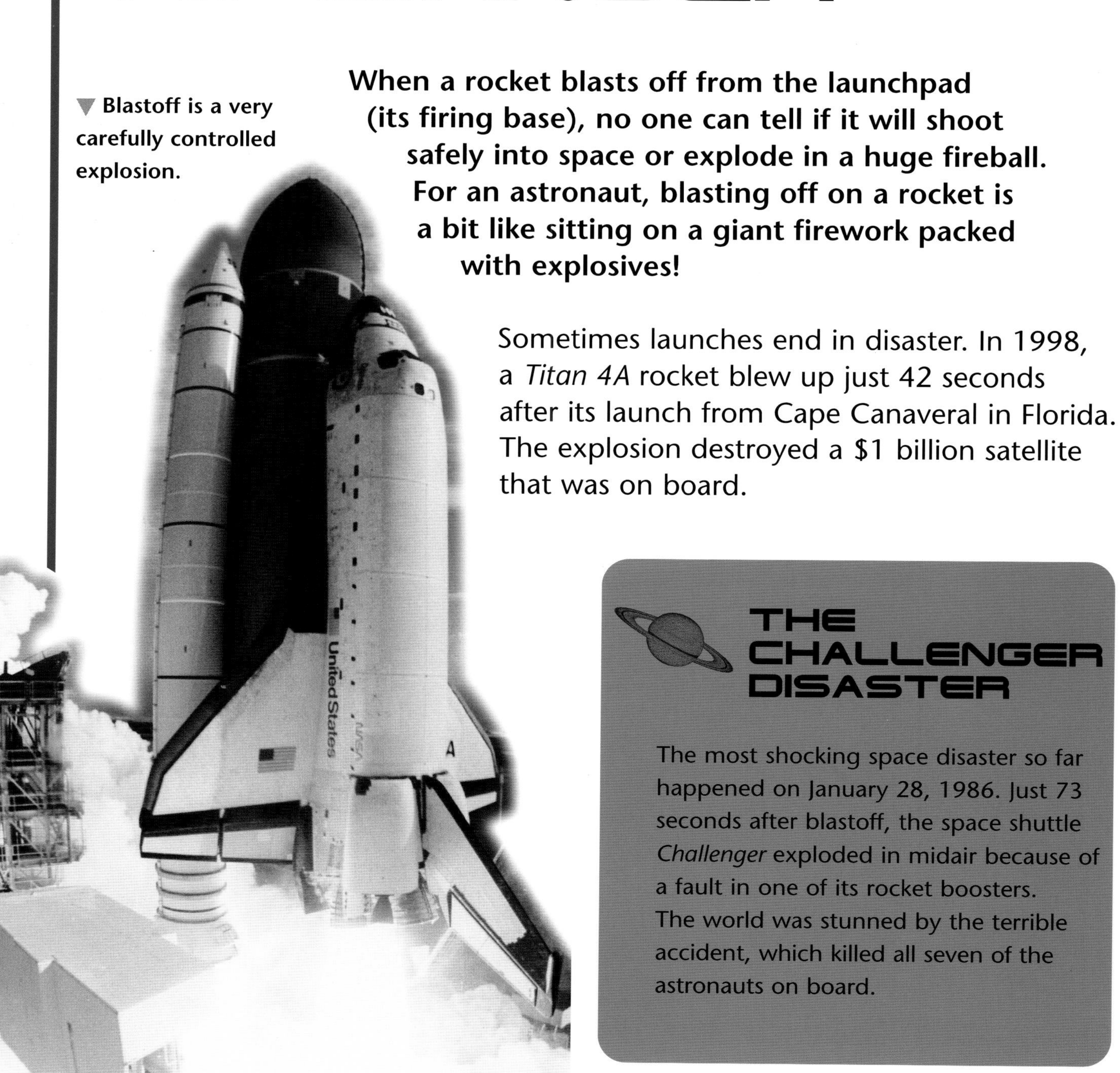

When a rocket blasts off from the launchpad (its firing base), no one can tell if it will shoot safely into space or explode in a huge fireball. For an astronaut, blasting off on a rocket is a bit like sitting on a giant firework packed with explosives!

Sometimes launches end in disaster. In 1998, a *Titan 4A* rocket blew up just 42 seconds after its launch from Cape Canaveral in Florida. The explosion destroyed a $1 billion satellite that was on board.

THE CHALLENGER DISASTER

The most shocking space disaster so far happened on January 28, 1986. Just 73 seconds after blastoff, the space shuttle *Challenger* exploded in midair because of a fault in one of its rocket boosters. The world was stunned by the terrible accident, which killed all seven of the astronauts on board.

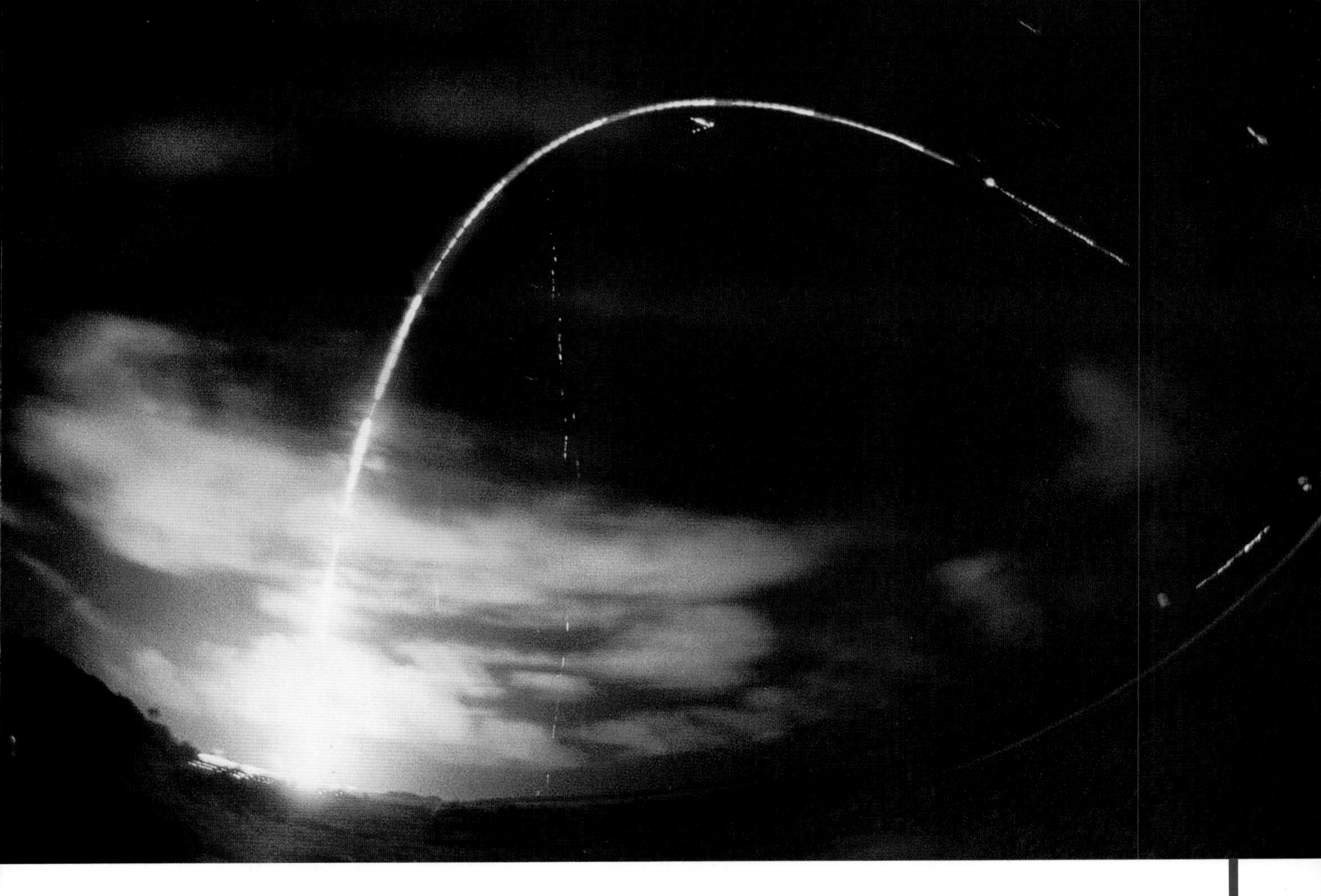

▲ **A European Ariane rocket blazes a trail as it lifts off into the sky.**

Two years earlier, a faulty computer program caused the European *Ariane 5* rocket to go off course. The rocket was worth $500 million and had taken ten years to develop. It had to be destroyed by people on the ground just 40 seconds after blastoff.

▶ **This satellite dish is used to control Ariane rockets from Earth.**

Stepping into Space

▼ Space suits are very special machines designed to keep astronauts alive.

Inside a rocket, astronauts are protected from any dangers. But sometimes they have to go outside the spacecraft to explore space or to make repairs.

Stepping outside a spacecraft is called a "spacewalk." The astronauts look as if they are floating in space, but they are usually tied to their craft by long support lines called umbilical cords. Astronauts wear space suits to protect their bodies from the incredible heat and cold, and from tiny meteoroids (flying space rocks) that may crash into them. Large helmets like goldfish bowls provide them with a mix of oxygen and nitrogen. Gold visors stop the Sun's dangerous rays from damaging their eyes.

In 1965, the Russian astronaut Alexei Leonov became the first person to walk in space. The next year, American astronaut Edwin "Buzz" Aldrin made a spacewalk lasting two hours, 20 minutes.

MAKING SPACE SAFER

The worst thing that could happen during a spacewalk would be for an astronaut's umbilical cord to break. The astronaut might drift off into space, never to be seen again. Fortunately, NASA (the U.S. space organization) has now developed a special kind of jetpack called SAFER. If an accident happens, the astronaut can use this to fly back safely to the spacecraft.

▼ Preparing for a disaster is part of every astronaut's training. Here, two astronauts practice the SAFER test during a spacewalk.

Lost in Space

Spacecraft are sometimes controlled by the astronauts on board and sometimes by mission control, the control center on Earth. When contact between mission control and a spacecraft is lost, the results can be dramatic.

Space probes are spacecraft with no astronauts on board. They are sent out to explore the universe. In 1999, NASA lost a $125 million probe called the *Mars Climate Orbiter.* This happened because of a simple mistake.

The NASA scientists had been using metric measurements (meters), while the satellite's builders had used standard measure (feet and inches). As a result, the satellite was steered too close to

◀ Expensive spacecraft sometimes disappear into space, never to be seen again.

Mars and burned up in the Martian atmosphere!

The team that steered the craft failed to follow NASA's rules. They also did not understand this spacecraft and how it was different from others that they were working on.

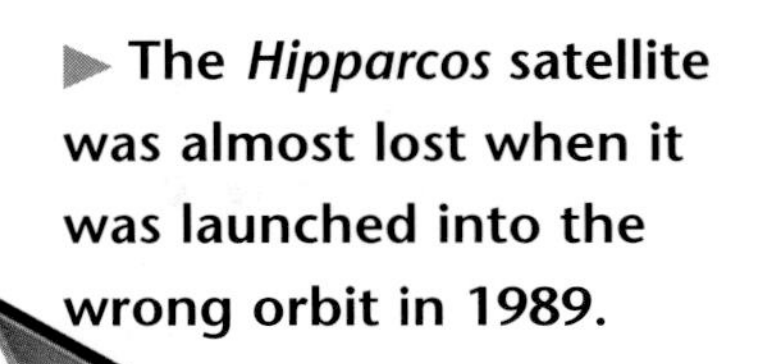

The *Hipparcos* satellite was almost lost when it was launched into the wrong orbit in 1989.

Scientists monitor, or check, spacecraft from their screens at mission control.

STRANDED ON THE MOON?

When the Apollo astronauts set off for the Moon in 1969, some people worried they might never return. America's president, Richard Nixon, had prepared a speech to read out to the world if Neil Armstrong and his colleagues became stuck on the Moon. Fortunately, the mission was a success and the astronauts returned home safely.

Big Booms

Most of the space dramas that we hear about happen when astronauts set out to explore the universe. But some very dramatic things happen naturally in space. These include the "big booms" caused by meteors and solar flares.

Meteors are lumps of rock that travel through space. When a meteor hits Earth, it can make a massive pit, or crater. One of the best known meteor craters on Earth is the Barringer Crater in Arizona. It is thought to have been made at least 50,000 years ago.

Some scientists believe the dinosaurs became extinct on Earth when a giant meteor crashed into our planet 65 million years ago. The chances of this happening again are very small, because the universe is such a big place.

▼ The Barringer Crater in Arizona is 2,625 feet (800 m) wide and 656 feet (200 m) deep.

THE CRASHING COMET

In July 1994, more than 20 pieces of a comet called Shoemaker-Levy 9 smashed into the planet Jupiter. Astronomers worked out that the explosions were bigger than those caused by a nuclear bomb.

◀ Comet Shoemaker-Levy 9 created a huge fireball when it crashed into Jupiter in 1994.

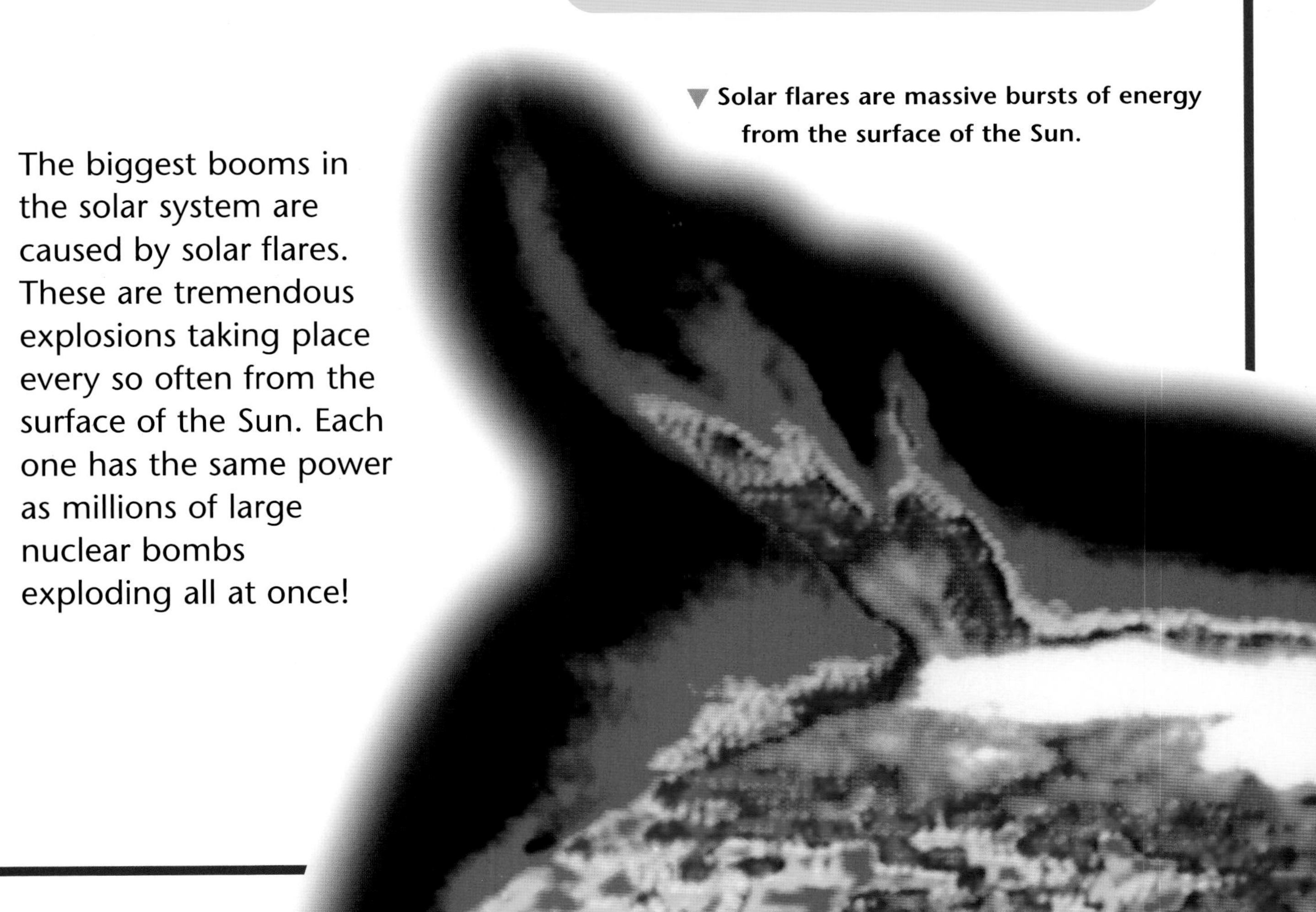

▼ Solar flares are massive bursts of energy from the surface of the Sun.

The biggest booms in the solar system are caused by solar flares. These are tremendous explosions taking place every so often from the surface of the Sun. Each one has the same power as millions of large nuclear bombs exploding all at once!

DANGEROUS DOCKINGS

▼ Russian and American astronauts met in space for the first time in July 1975, when their Apollo and Soyuz spacecrafts docked.

Spacecraft need to join together, or dock, with the space station, to allow people to get on and off the station. Docking is difficult and dangerous, and sometimes things go wrong.

The very first docking of two spacecraft happened in 1966. The rocket *Gemini 8* was piloted by astronaut Neil Armstrong. It was supposed to dock with an unmanned craft called the GATV (Gemini Agena Target Vehicle). It took six hours for the craft to meet up, and then disaster struck.

▼ This photograph of the GATV is what Neil Armstrong saw from *Gemini 8* in 1966, just before the two craft docked.

A fuel leak made the two docked spacecraft spin around and around. Armstrong decided to separate them, but this only made things worse. Soon, his spacecraft was spinning so fast, once a second, that Armstrong and his co-pilot could no longer see properly. Armstrong had no choice. He had to end the mission. *Gemini 8* made an emergency splashdown in the Pacific Ocean, and the astronauts were picked up safely.

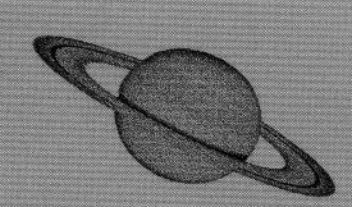

A HIT FOR MIR

Russia's space station *Mir* provided lots of drama during its 15 years in space. One of the biggest was in 1997, when a robot-controlled cargo ship crashed into *Mir* during docking. The ship weighed over 7 tons. The Russian cosmonauts and U.S. astronauts on *Mir* had just half an hour to fix the damage. Later, they had to carry out a dangerous spacewalk to turn *Mir*'s solar panels toward the Sun, so the craft could get power again.

Close Encounters

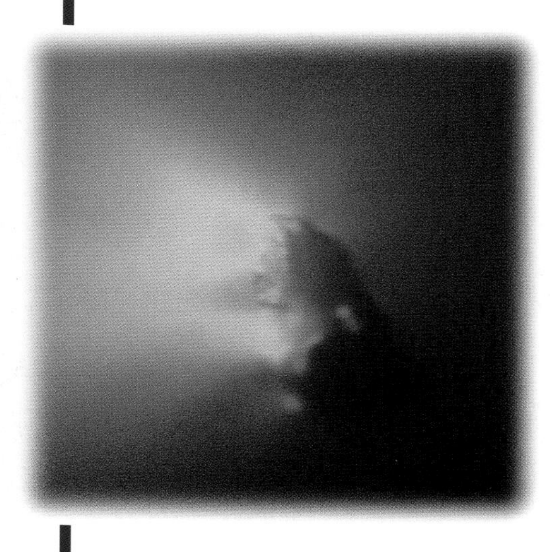

▲ Halley's comet photographed by *Giotto.*

Space probes are often sent on missions to find out more about comets or asteroids. Comets are sometimes called dirty snowballs because they are made up of ice and dust. They have a center part called the nucleus and a long tail of dust and gas.

Scientists found out much more about comets by sending a space probe called *Giotto* to meet the most famous comet of all, Halley's comet. This close encounter happened 58 million miles (93 million km) from Earth in March 1986. *Giotto* passed within 376 miles (605 km) of the comet's nucleus, the leading "snowball," and was knocked around by the dusty wind from the comet as it passed by.

◀ *Giotto* was traveling faster than a bullet when it passed Halley's comet in 1986.

▲ **The NEAR probe prepares for its historic landing on the Eros asteroid in 2001.**

NEAR AND FAR

On February 12, 2001, a small probe called *Near Earth Asteroid Rendezvous* (NEAR) *Shoemaker* managed to land on top of an asteroid called Eros after circling around it for about a year. This dramatic meeting happened 120 million miles (193 million km) from Earth.

Asteroids are small planets that circle the Sun in a broad band between the planets Mars and Jupiter. In 1991, the space probe *Galileo* made the first close-up study of an asteroid called Gaspra during its voyage to study Jupiter. *Galileo* and Gaspra flew past one another at 18,020 miles per hour (29,000 kmh). (That's over ten times faster than the top speed of a jet fighter!)

Daring Repairs

Spacecraft are incredibly complicated machines. When things go wrong with them, astronauts may need to risk their lives to make repairs. But even when the repairs have been made, other things may go wrong!

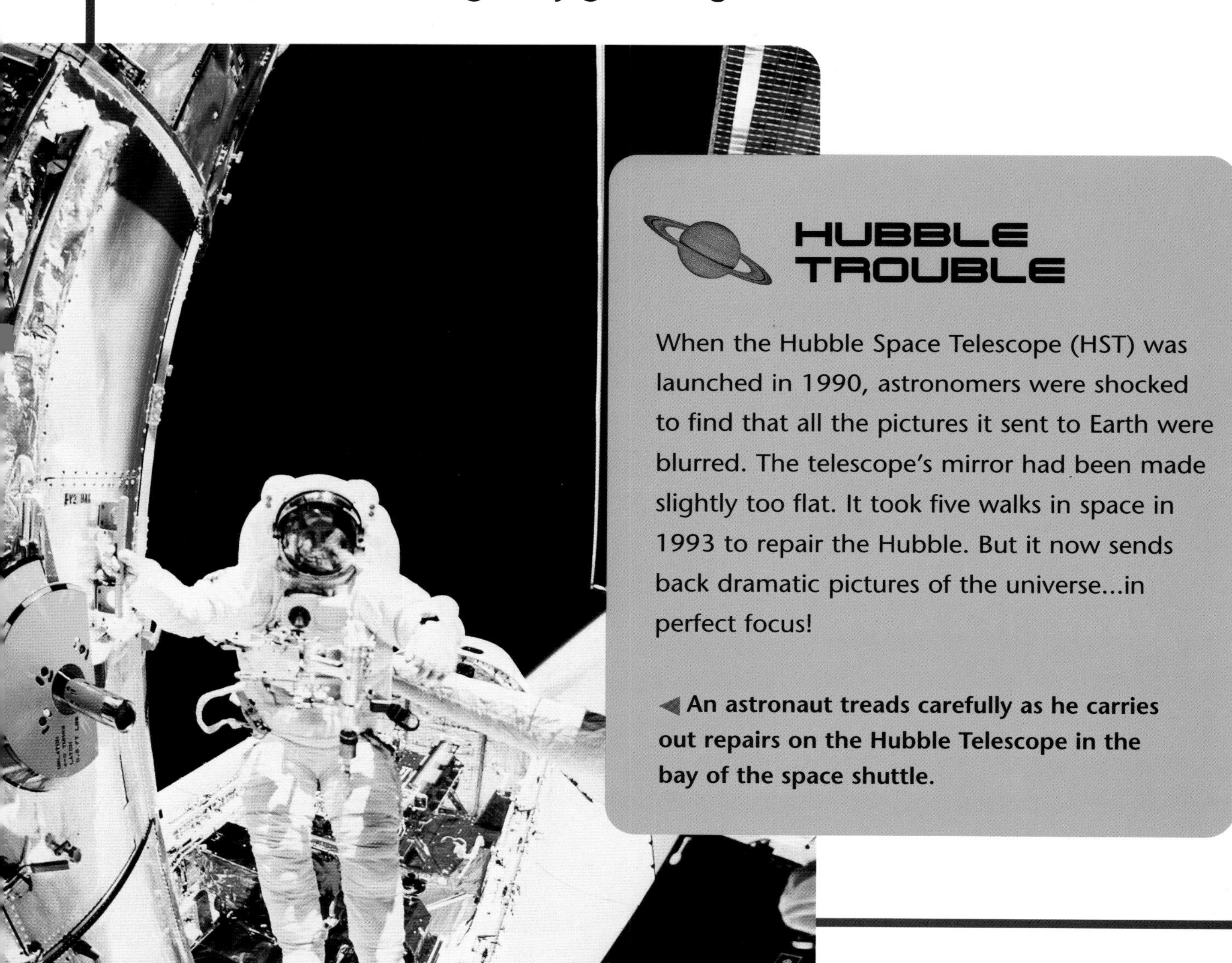

Hubble Trouble

When the Hubble Space Telescope (HST) was launched in 1990, astronomers were shocked to find that all the pictures it sent to Earth were blurred. The telescope's mirror had been made slightly too flat. It took five walks in space in 1993 to repair the Hubble. But it now sends back dramatic pictures of the universe...in perfect focus!

◀ **An astronaut treads carefully as he carries out repairs on the Hubble Telescope in the bay of the space shuttle.**

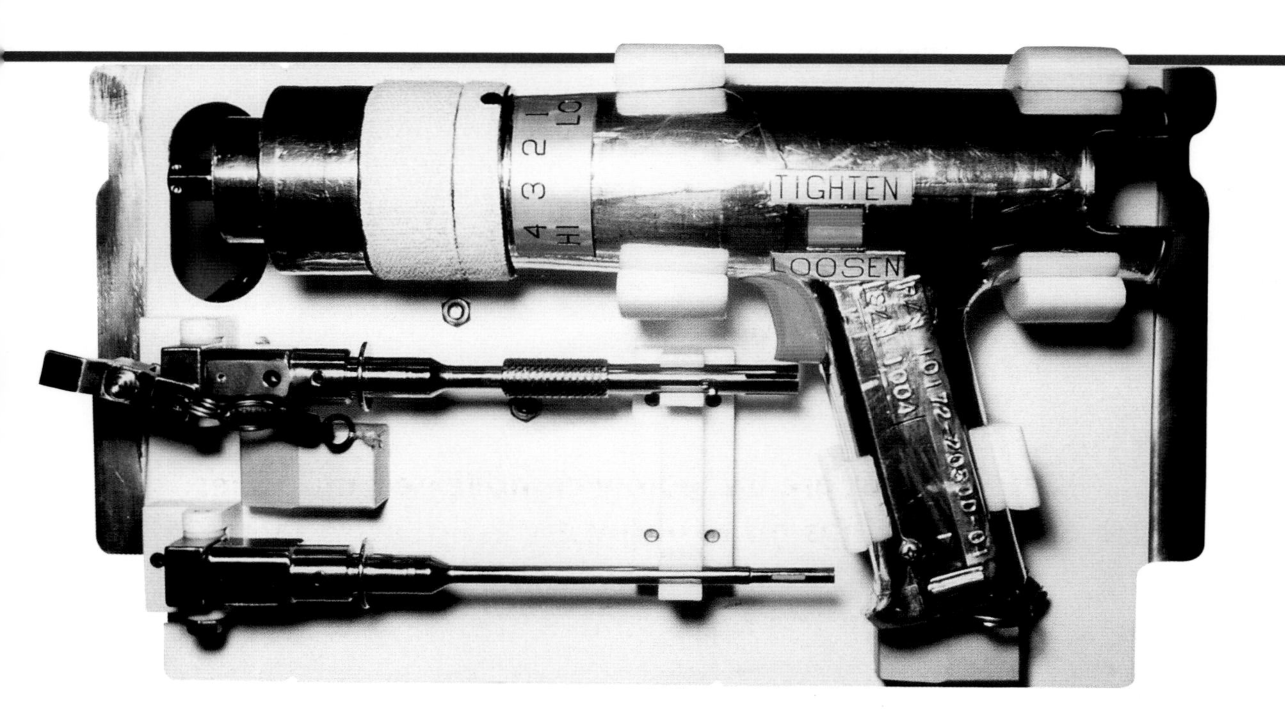

▲ **Even space repairers need their tools. This is a battery-powered screwdriver used by astronauts.**

NASA's first space station, *Skylab 1*, was damaged when it took off in May 1973. As it flew into orbit, an important shield was ripped away from the craft. This had been designed to protect *Skylab* from the heat of the Sun and from meteoroids. When the shield broke away, it took one of *Skylab's* solar panels with it and pushed the other one out of position. Solar panels are long flat panels that use the Sun's energy to make power. The craft was left without power.

Three astronauts took off in another rocket to make repairs. One of them leaned out of an open hatch and used a large hook to pull *Skylab's* solar panel back into position. Next, the astronauts put up a giant umbrella over the outside of *Skylab* to keep it from being damaged by the Sun.

◄ **The first repair of a satellite in space took place in 1984.**

Risky Rescues

When accidents happen in space, the most important thing is to get the astronauts safely back to Earth. All astronauts prepare for emergencies as part of their training. Even so, space rescues are always difficult and dangerous.

American astronaut Virgil "Gus" Grissom almost lost his life in July 1961 when his Mercury capsule splashed down in the Atlantic Ocean. The hatch flew open too soon, causing the capsule to fill with water and then sink. Grissom almost drowned before being rescued by a helicopter.

▲ **A U.S. Marine helicopter tries to rescue Gus Grissom from the Atlantic Ocean.**

► **Astronauts train in giant swimming pools on Earth. Here, they are preparing for an emergency on the International Space Station.**

Sometimes accidents happen too quickly for astronauts to be rescued. In 1971, the Russian *Soyuz 11* spacecraft was on its way back to Earth when something went wrong. An air valve blew open, causing all the air in the capsule to be sucked out. The three cosmonauts on board were killed by the lack of oxygen as their craft flew back to Earth.

APOLLO RETURNS

Disaster struck the *Apollo 13* mission to the Moon in April 1970. Halfway to the Moon, a tank of oxygen exploded. *Apollo 13* lost power, and the astronauts had to be rescued by mission control at Houston, Texas. Millions of people around the world watched the rescue on television and prayed for the astronauts' safe return. The rescue was a success, and *Apollo 13* splashed down safely in the South Pacific Ocean four days later.

▶ **Astronaut John Swigert is rescued from the ocean by a helicopter, following the *Apollo 13* splashdown.**

Unhappy Landings

▲ Large pieces of Skylab crashed down in Australia in 1979.

Spacecraft that take off from Earth do not always come back again. Some stay in space forever. Others burn up in Earth's atmosphere on the way back. And some come crashing back to Earth when they are least expected!

Many people were worried when NASA decided to bring its *Skylab 1* space station back to Earth in 1979. NASA was unable to tell where the station would crash land. Fortunately, most of the craft burned up as it passed back into Earth's atmosphere, but some did come back to Earth.

▼ When the space shuttle returns to Earth, a large parachute fired from the back slows it down and makes sure it has a happy landing.

The Russian space station *Mir* had an unhappy landing, when it burst into flames as it returned to Earth in March 2001.

In April 1967, cosmonaut Vladimir Komarov became the first man to be killed on a space flight. The parachutes on his *Soyuz 1* spacecraft tangled up and did not open. Instead of slowing down, *Soyuz 1* raced toward Earth and crashed at a speed of 90 miles per hour (145 kmh). Komarov was killed instantly.

MIR DISASTER

When the space station *Mir* returned to Earth in March 2001, some of it had burned up in the atmosphere. Other parts had crashed in the ocean. But if *Mir*'s engines had failed, it could have landed on Japan. The Russian government was so worried *Mir* would crash on a city that they took out a $200 million insurance policy to cover them against any damage.

Russian cosmonaut Vladimir Komarov, who was killed on his return to Earth in 1967.

FUTURE DRAMAS

Space travel has always been exciting and dangerous. In the future, astronauts may travel farther and farther from Earth. And who knows what they may encounter—bits of space junk, new galaxies, or even aliens!

People have put thousands of objects into space since 1957. Over half of them have since come back to Earth. But quite a lot of "space junk" still remains in orbit. Space junk moves very fast, and it can cause a great deal of damage when it hits something.

▼ **The International Space Station has taken years to plan and build. Could it really be destroyed within just a few minutes by space junk?**

One danger is that some of the junk could fall back to Earth and kill people. The other danger is that it could stay up in space and crash into satellites or spacecraft. No one on Earth has yet been killed or injured by falling space junk. But because there is so much junk in space the chances of it hitting the new International Space Station are high. Could the next big space drama be caused by space junk?

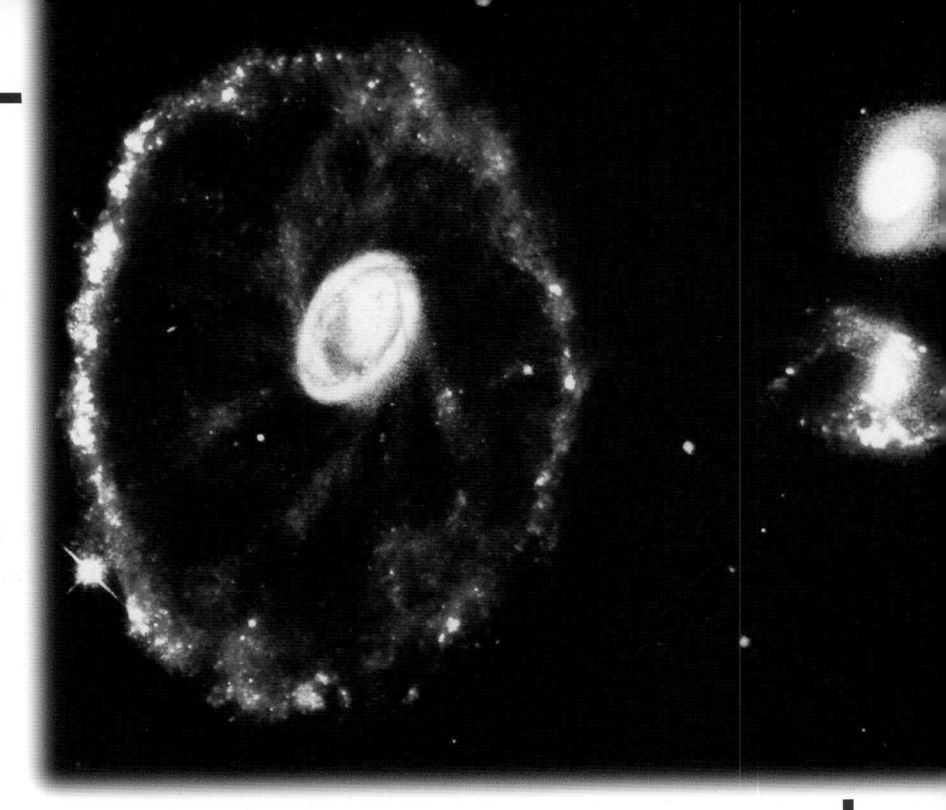

▲ The cartwheel galaxy is a new discovery. Discoveries like this one show us how little we know about space.

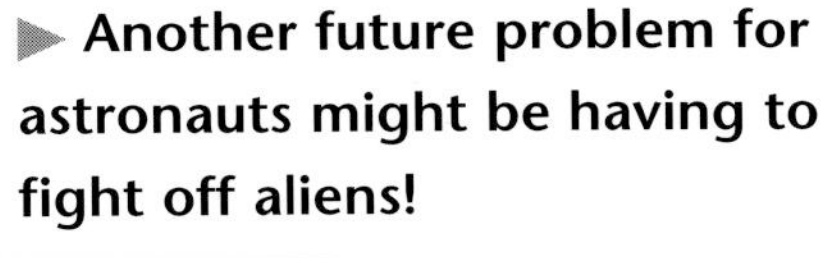

▶ Another future problem for astronauts might be having to fight off aliens!

MARTIAN MADNESS?

Astronauts who take part in a future trip to Mars may be at great danger from a type of space radiation called cosmic rays. These rays are so powerful that they can damage the human brain and stop a person from ever being able to think clearly again. Researchers are now trying to develop computer systems that can check if astronauts are suffering from the effects of cosmic rays.

Space Drama Facts

Blastoff and splashdown dramas
When a *Titan IV* rocket exploded on liftoff in 1998, the noise of the blast set off car alarms over 10 miles (16 km) away.

Since *Sputnik 1* took off in 1957, 26,643 objects have gone into orbit around Earth. Some are satellites and spacecraft. Some are bits of space junk. Since then 18,000 of these objects have come back to Earth.

A *Mercury-Atlas 7* rocket missed its landing spot in May 1962. A rocket failure meant that it overshot the landing zone by 249 miles (400 km).

Computer failure
Two NASA Phobos space probes were lost in 1988. *Phobos 1* moved too far from the Sun, so its solar batteries died. Control of *Phobos 2* was lost when its on-board computer failed.

Sun cycle
The Sun goes through a cycle that comes to a peak every 11 years, with lots of extra sun spots, rays, and other activity. This can affect our satellites, radios, cell phones, and electricity supplies here on Earth. The last peak was in 2000.

◀ **The *Mars Polar Lander* reached Mars in 1999. Then communication was lost, and the mission abandoned.**

◀ **The space shuttle *Challenger* exploding in 1986.**

Death in space
A new Russian rocket was being tested in Kazakhstan in October 1960. Field Marshal Mitrofan Nedelin arrived to watch the launch. Problems developed before the test, but Nedelin ordered it to go on. Then, the engine caught fire and exploded. The explosion created a poisonous cloud that damaged the lungs of people nearby. At least 91 people were killed.

On January 27, 1967, a fire broke out during a launch test of the *Apollo 1* spacecraft. Three American astronauts, Virgil "Gus" Grissom, Edward White, and Roger Chaffee, were killed.

The space shuttle *Challenger* exploded just 73 seconds after liftoff. All seven of the astronauts on board were killed. *Challenger* was one of several space shuttles used by NASA. It had flown nine times before. The disaster was caused by the failure of a seal called an "O-ring" on one of the rocket boosters.

Space Drama Words

asteroid (AS-tuh-roid)
A lump of rock or very small planet that orbits, or moves, around the Sun.

astronomer (uh-STRON-uh-mur)
A person who studies the stars, planets, and space.

blastoff (BLAST-awf)
The launching of a rocket into space (liftoff).

comet (KOM-it)
A small body made of ice and rock that orbits the Sun. A comet has a main part and a long bright tail.

cosmic rays (KOZ-mik RAYZ)
This is dangerous radiation that travels through space.

cosmonaut (KOZ-muh-nawt)
A Russian astronaut.

docking (DOK-ing)
When two spacecraft join together in space.

launchpad (LAWNCH-PAD)
The place where a rocket blasts off into space.

meteoroid (MEE-tee-uh-roid)
A small chunk of rock in space.

mission (MISH-uhn)
A voyage to space.

mission control (MISH-uhn kuhn-TROHL)
The job of mission control is to watch over and control a space mission from its base on Earth.

◀ **The long, bright tail of Halley's comet.**

the astronaut in different directions.

solar panel (SOH-lur PAN-uhl)
A device that changes sunlight into electricity. Solar panels are often used to power spacecraft.

space junk (JUHNGK)
Broken pieces of rockets and old satellites that stay up in space.

space probe (prohb)
A spacecraft with no astronauts on board. Space probes are controlled by on-board computers and mission control back on Earth.

space station (STAY-shuhn)
A spacecraft that remains in space for a long period of time. Astronauts can live and work in a space station for months or even years.

spacewalk (WAWK)
When astronauts go outside their spacecraft, while in space, to explore space or make repairs.

splashdown
When a spacecraft returns to Earth and lands in the ocean.

umbilical cord (uhm-BIL-uh-kuhl)
The safety line that connects an astronaut to a spacecraft during a spacewalk.

NASA (NAS-suh)
The National Aeronautics and Space Administration, which organizes space exploration on behalf of the U.S. government.

nuclear reactor (NOO-klee-ur ree-AK-tur)
A device that provides power for a spacecraft by using nuclear energy. Nuclear energy is released when certain atoms in the fuel are split apart.

nucleus (NOO-klee-uhss)
The main, leading part of a comet, made of ice and rock.

SAFER jetpack (SAYF-ur JET-PAK)
This stands for Simplified Aid for EVA (extra-vehicular activity) Rescue. It is attached to the base of a normal jetpack. Its jets provide thrust to move

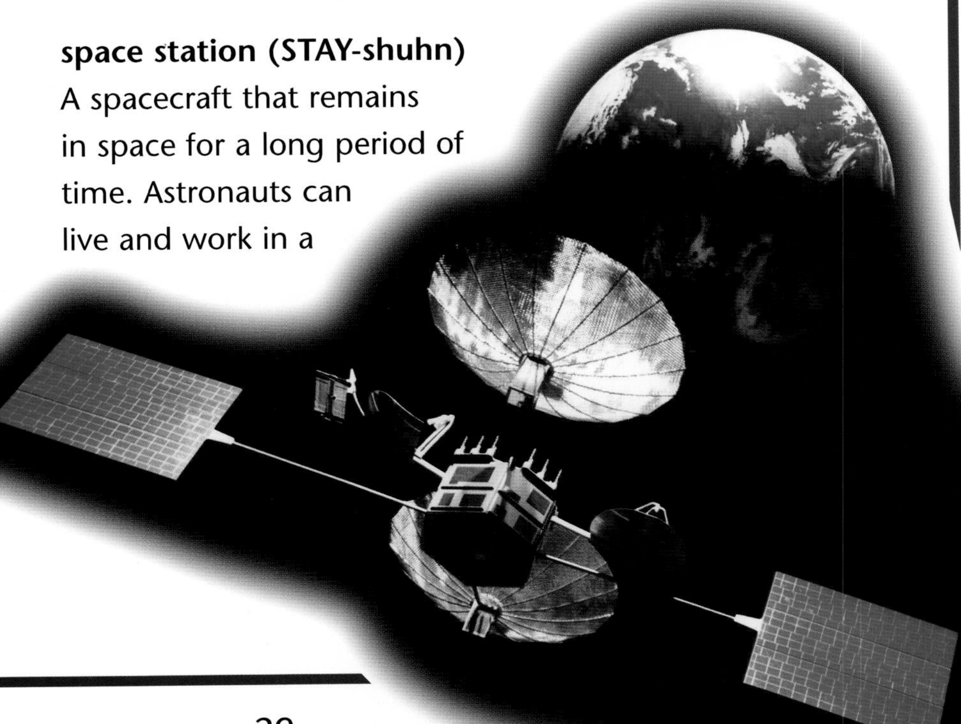

▶ The large, square solar panels on this satellite produce electricity to keep it moving through space.

SPACE DRAMA PROJECTS

VISIT A SPACE BASE

NASA has a number of space bases around the United States and some of them are open to the public. If you go on vacation, perhaps you could take a trip to a space base. You can walk around the base, look at real-life rockets, and find out more about how hard it is to put a rocket into space.

You can find a list of NASA's bases on the Internet at: www.nasa.gov/visitors.html

NASA also tells people about the dates of rocket launches so visitors can watch. You can find more information at: www.nasa.gov/see_a_launch.html

STUCK ON THE MOON?

Imagine you are an astronaut, and you have just taken off from the Moon. You are about to come back to Earth when there is a huge explosion. Your spacecraft seems to be stuck on the Moon! Write a story about your adventure. Can you repair your rocket? How? What items can you find in the rocket to help you? Perhaps you can find some useful spare parts on the Moon? Explain how you get back to Earth.

DRAW A SPACE DRAMA

Pick one of the dramatic stories in this book, and try to draw a picture of it. Maybe you could draw the spinning spacecraft *Gemini 8*, or an astronaut taking a walk out into space. Or you could imagine your own drama about the future and record what happened in pictures.

SPACE DRAMAS ON THE WEB

If you have access to the Internet, you will be able to track down information about space including photos of stars, planets, galaxies, spacecraft, and information about space missions.

NASA for Kids
www.nasa.gov/kids.html
This site provides information about the stars and planets, rockets, pioneers, and astronauts. You can play games and puzzles and take part in on-line activities.

The Space Place
spaceplace.jpl.nasa.gov
A great site with lots of things to make and do and some amazing space facts.

Space kids
www.spacekids.com
An excellent site with photos, videos, facts, games, and help with projects for homework.

NOVA Online: Terror in Space
www.pbs.org/wgbh/nova/mir/textindex.html
This site tells you about some of the biggest disasters in the history of space flight.

▲ **Tourists watch an astronaut at the Kennedy Space Center in Florida.**

NEAR voyage Earth to Eros
near.jhuapl.edu/Voyage/
This site takes you step-by-step on a voyage from Earth to the Eros asteroid.

You can search for other space sites using any search engine. Try using specific search phrases such as the name of a mission, astronaut, or spacecraft.

INDEX